Prove It

Diagnose System Failures Without Guessing

Sgt. Cognition

Cognition Admission

Prove It: Diagnose System Failures Without Guessing

Published by Cognition Admission

www.sgtcognition.com

ISBN: 978-1-970905-04-5

First Edition, 2026

Table of Contents

Prelude

So you want to be an HVAC technician.

Good news, you're sucking air. Bad news, not everyone gets a uniform.

Let me be straight with you before we go any further. This trade is survival of the fittest. The men who last in this industry run on caffeine and a beef stick and still show up with a smirk and a tool bag. Not because the job got easier. Because they got harder.

I have spent over twenty years as a lead senior service technician. The majority of that time was spent walking into systems that someone else already touched. Someone who was confident. Someone who was sure. Someone who left before the callback arrived and let the next technician inherit the mess.

That next technician was usually me.

480 volts does not negotiate with what you think you know. A roof in January does not care about your comfort level. The family standing in their living room at 11 p.m. in August with no air conditioning is not interested in your excuses. The road between you and that call does not adjust for weather, fatigue, or the bad decision you made the night before.

This trade will test every part of you before it gives you anything back.

There is no shame in recognizing that right now and walking away. This is the moment to make that call. Nobody will fault you for it.

But if you want to shadow me, if you want to learn the method that gets a technician through twenty years of calls, callbacks, bad systems, and worse situations, you are welcome to ride along.

Just understand what you are riding into.

This is not for the technician with a search engine education and a gut feeling. This is for the man who shows up. Every single day. Despite the weather, despite the discomfort, despite knowing exactly how ugly it's going to get before he turns the key.

He shows up anyway. As the professional on site.

Everyone else can kick rocks.

Let's get into it.

The method you are about to learn is not theory. It is the difference between a technician who proves his work and one who guesses at it. One of those men goes home at the end of the day knowing the job is finished. The other one waits for his phone to ring.

This is how you stop guessing.

Chapter 1: What You Walked Into

HVAC has a culture that is not unique but it is built different.

Let me explain.

Experience is currency in this trade. Not job titles. I have worked under service managers with zero field experience who spoke exclusively in numbers. So the question becomes, how do you follow a leader who has never diagnosed a system in his life?

You follow him because you understand his job is not your job.

The manager sees numbers. That is his function. He keeps the lights on, manages the logistics, and handles the problems you never see from the field. He is not there to fix your diagnosis. That is why you were hired. Boots on the ground. His world and your world operate on different tracks and that is not a flaw in the system. That is the system.

Here is what you need to understand about that relationship. Do not go to your manager with problems alone. Go to him with a problem and a solution. Learn that the easy way by reading it today or the hard way dodging a cup of stale cold coffee. Either way you will learn it.

Your lead technician is the man you bring your technical problems to. He answers your field questions. Your manager gets involved when it becomes a company problem. Know the difference and use it correctly.

Say you catch a rooftop unit on fire. This can happen. This has happened. First call is to your manager. "Hey boss, the fire chief would like a word with you." Second call is to your lead or senior technician. That is where you explain what happened. "I crossed a red and white wire and briefly made eye contact with the afterlife. Can you tell me why?" That tech will be on site before you finish the sentence.

Why? Because nobody in this trade wants to see their brother get hurt. It also extends their on-call coverage if you are not available. Joking. Mostly.

Much like the military, there is a ranking system in this trade. Learn it. Use it correctly.

Now here is where it gets complicated.

The experienced tech is invaluable. The man who hates boilers but can give you a complete sequence of operations with amp draws after a twelve hour shift, half asleep, that man is a resource you do not waste. His knowledge is real and it was earned the hard way.

But experience carries bad habits the same way it carries good ones.

A senior tech tells you to jump R and W. That may be the correct move. He may know this system, may have been on site before, may know exactly what is needed to get you moving. Fine. But you still do not know. You felt the pressure to get the call closed and you got an answer that fixed the problem.

What did you learn? How to use a jumper.

That is not enough.

When information comes in, filter it. Ask why R and W matters. Find the schematic if one exists. Trace the sequence before you move. Understand the answer before you apply it. Because at the end of the day your name goes on that service ticket. Not his. Yours. He was busy when you called him and he is busy right now solving problems the rest of the company cannot handle.

Lead and senior technicians are not sitting by the phone waiting for your questions. They are standing in front of hostile customers, systems that have been misdiagnosed three times, and a company reputation they are personally holding together. Then you call.

Think before you call. When you get an answer make it your own. Understand it completely. Call that same tech the next day with the same question and you may receive a response that is not appropriate for a professional environment. We work on offices. We do not sit in them. But that standard still applies.

The reputation you do not want is simple. "This guy asks the same questions twice and learns nothing."

Do your own homework. Ask questions based on that homework. If you need to rent a senior tech's brain every time you hit a wall, fast food has a more appropriate career path for you.

This industry is direct. Be your own man. Do your own work. Think for yourself. Every senior tech in this trade loves to give clarification. Nobody can give you the willingness to study on your own. That part is yours and nobody can carry it for you.

Neglect it and you build a bad reputation. Bad reputation means callbacks. Callbacks mean the customer stops trusting you. The customer stops trusting you and your company has a problem with your name attached to it.

Survive the culture. Understand the hierarchy. Do your own work.

What comes next is even more important.

Your safety. In your truck and in the field.

Chapter 2: The Truck

Let me say this plainly and clearly before anything else.

The company truck is not your truck.

I am exhausted from watching technicians develop the mindset that because they were issued a vehicle it belongs to them. Negative. That vehicle is company property and is to be treated as such. No stickers on the exterior. No smoking lounge in the back. No speeding. No attempting to run over fast moving squirrels for entertainment.

You are driving a billboard with the company phone number on the side. Every traffic violation, every aggressive lane change, every close call on the highway is a public performance reviewed by anyone within eyeshot. Any technician at any time can be assigned the vehicle you are currently driving. It belongs to the company and they take their fleet seriously. They take driving records seriously. They take fleet maintenance seriously.

So you better be serious. Otherwise you get a serious pink slip.

Every day before your shift starts that truck gets inspected. Key in the ignition, lights on, four ways flashing, full walk around, deep dive inside. All fluids checked. Tires. Wipers. Everything. Treat that vehicle like it is your primary defense against the outside world getting you home at the end of the day. Because that is exactly what it is.

As an HVAC technician you will spend significant time behind the wheel. Hours on the road, sometimes in desolate areas, sometimes in conditions that have no interest in cooperating with your schedule. We call it windshield time. Some techs love it. Some hate it. All of them have to live with it.

Your truck takes serious abuse. Daily mileage, road debris, snow, rain, and everything else mother nature decides to contribute. Most techs assume call to call that the truck is going to do its job.

Until it doesn't.

Then you are stranded somewhere waiting for another technician to come collect you. Hours pass. You finally get picked up, brought back to the office, handed the spare truck, which is never a good truck, and then you go finish your calls for the day. You thought a breakdown meant a lighter day? It means a longer day with a worse truck and an attitude you did not start with.

All of that is avoidable with a five minute walk around and the appropriate checks before you leave the yard. Report any deficiency to your chain of command. If your company does not have a checklist for this, build one based on the vehicle and what the manufacturer recommends. You will not find one in this book. Do your own homework.

What you will find here is this.

Everything in the back of that truck must be secured and compliant. No refrigerant jugs rolling around on the floor. They go in a rack. They get secured. Every single item in that vehicle needs to answer some questions before the doors close.

What happens to this if I get hit at 70 miles per hour?

What happens if this truck rolls?

Does this item become shrapnel in a steel cage?

I do not want to hear about the cab guard protecting the driver. I don't have to guess at how well that guard holds up against a loose black iron pipe traveling at speed.

Against a loose acetylene tank. Against an open jug of used oil with nowhere to go but forward.

I have seen what happens.

An installer flipped his vehicle at high speed and the truck happened to land into someone's living room. First thing I noticed was a black iron pipe that went straight through the cab guard and into the headrest of the passenger seat. Mushroomed like a bullet. The second thing I noticed was screws that blew through the cage gaps and out the windows into the structure. The third thing I noticed was thc empty cough syrup bottles on the floorboard.

So did the homeowner.

Nobody was physically injured in that crash except the installer's career and the company's bank account. But the margin between that outcome and something permanent was thin. Everything in that truck becomes a weapon in seconds. Secure it. No exceptions.

Drive the company truck like it belongs to someone else and like everyone around you is recording your speed. More importantly make it safe for any technician who may operate that vehicle after you. How would you feel if a tech swapped into your assigned truck for the day and a can of grease blaster you left on the floor rolled under his brake pedal?

Nobody wants that outcome. Including you.

Five to ten minutes every morning eliminates the variable. You cannot control the discipline of everyone else on the road. You can absolutely control yours. Eliminate what you can eliminate and account for what is yours to account for.

Truck inspection is as important as equipment verification. That comes later in this book.

You survived the truck. You made it to the call.

Now it is time to talk about the next problem entirely. The customer. Keep reading. It is about to get spicy.

Chapter 3: The Customer

If you have ever heard someone say they trust animals more than people, they may have been an HVAC technician in a previous life.

Anyone who has spent time in customer service rolls their eyes at least twice a day. But HVAC is different from standing behind a counter. When you are in someone's home you are not dealing with a customer in the traditional sense.

You are dealing with your boss.

That castle already has a king. What remains to be seen is whether you are added as a jester or an advisor. History has been pretty clear about what happens to jesters.

Never forget that when you are in someone's home you are a guest. You are not their buddy. You do not refer to them as such. The highest level of respect gets established at the door before a single tool comes out of the bag. Leave your problems in the truck. Whatever is going on in your life is irrelevant the moment you knock on that door. This is no longer about you.

Understand what you are walking into.

A man whose heat went out had to swallow his pride and call for help. His queen is quietly reconsidering her life choices. His kids think he can fix anything and are watching to find out if that is still true. He is embarrassed, skeptical, and already on edge before you arrive. Would you not be?

Leave your ego in the truck with your problems. You will not need either one inside.

Once you cross that threshold the sensory overload begins. You are either the hero at the start or the villain at the end, sometimes both within the same visit. Everyone is cheering when you walk in. Until they see the bill.

“This is outrageous. I could have flipped that emergency switch myself. I want this itemized.”

Classic. And as a junior technician that moment can start a stutter chain. “Well uh, it’s what the company charges, not really my fault, sorry.”

Wrong. That response is your fault. You are a professional. You diagnosed and resolved a problem and that is payable by any reasonable standard. Sometimes you just have to remind people of that.

Like this.

“I understand how you feel. Let’s take an honest look at what happened. You had no heat, you called, I found the emergency switch in the off position. You are absolutely right, you could have flipped it yourself if you had known. But ask yourself honestly, if I walked into your place of work tomorrow, could I expect the same discounted rate on your services? It costs time and money to resolve a problem. That is why there is a charge. I would trade you for a cookie if I could but my boss is diabetic and I do not smell anything baking. So here we are.”

Be respectful. Stand your ground. You are not a discount operation. Your company has financial obligations and if you are not pulling in the revenue required to sustain the operation you become overhead. Nobody wants to be overhead.

Now. Some people will dig their heels in regardless of how reasonable you are. They want what they want and they are not hearing anything else. You must recognize that line.

Without it you are pushing a customer toward paying something they have already decided they fundamentally disagree with, and you are doing it inside their castle.

Recognize when it is time to close your mouth, pack your gear, and contact the office.

Your manager handles the money disputes. That is his lane. Your job is to do right by the customer while you are in their home and represent every technician in your company who will ever knock on that door after you. One tech with an attitude demanding payment in someone's living room damages the reputation of every single person wearing that uniform. Do enough of that and your coworkers start looking at you sideways. You do not argue. You handle what you can, respect the client, and hand it off when it goes beyond your lane.

Keep your tone professional. Light when appropriate. Never mistake being approachable for being their drinking buddy. You are a professional, not a bar bouncer.

This is easy to say and genuinely difficult to do. Especially when the customer is angry or following you around like a shadow. I have had a customer rest his chin on my shoulder while I worked. Watching every move.

They exist. You will meet them. There is a method for handling them too.

Read the person before you react to them. Most hovering customers are not hostile. They are operating from one of three places. They want to learn enough that they never have to call you again. They genuinely want to be useful. Or they are simply curious about what is happening inside equipment they own and are responsible for.

Use that.

Invite the curious customer to hand you tools. Bring the angry customer down to see exactly what you are doing. Explain what you found in plain language, no trade terminology, no condescension. Speak to them like the adults they are even when that designation feels generous.

The best explanation connects your diagnosis to something the customer already understands.

Dog groomer? Perfect. "The filter in your system collected enough debris to trip a safety switch. It is similar to when you get a blow dryer too close to a dog and it shuts off from the heat. You wait and it resets. Same concept here. The difference is the component that tripped is bimetal, meaning it degrades a little each time it activates, the same way brake pads wear with use. That is why I am recommending replacement even though the system is currently running. I am happy to share my readings so you can see exactly what I found and reach your own conclusion."

That is how you deliver a diagnosis to someone who does not speak your language. Meet them where they are. Lose the jargon. Give them something real to hold onto.

Think about the last time you took a vehicle to a mechanic and knew nothing about what was wrong. How did you want that mechanic to explain why you just spent five hundred dollars? Apply that standard every single time.

People will always be a challenge in this trade. You will be lied to. You will be manipulated, sometimes with force and more often with kindness. One of the hardest parts of this job is navigating people inside spaces they own while holding knowledge they do not have.

Remember who holds the power. The owner of the equipment does. That is always true. Also remember that you are holding power too. Somewhere between 24 and

480 volts, it has been patiently waiting for you to stop paying attention.

The method that follows is not about the customer. It is not about the truck or the culture or the politics of the trade. It is about the system in front of you and whether you are disciplined enough to let it tell you what is wrong before you decide you already know.

Most technicians never get there. You still can.

Chapter 4: The Discipline of Diagnosis

There are two types of technicians. The one who proves his work and the one who guesses until something changes.

Both of them can leave with the equipment running. Only one of them can tell you why it runs.

Pattern recognition feels like expertise. You have seen something like this before. The symptoms connect to a past job. The brain starts building a conclusion before the system finishes showing you what it is doing. It feels efficient. It feels like experience doing exactly what it is supposed to do.

Then the system stops cooperating with the pattern.

There is a hard difference between recognizing a pattern and proving a cause. They look identical right up until the moment they do not. One brings the system back under control every time. The other works just often enough to build unearned confidence and then quietly schedules your return visit.

Confidence says I know what this is. Knowledge only exists after proof. Discipline is what forces the proof.

Speed without structure is faster guessing. When experience runs without discipline the brain takes shortcuts it does not announce. It fills in gaps before they are checked. It starts finishing the system's sentence before the system is done talking.

Every call has a fork. What seems likely based on experience versus what can be confirmed based on what the system is doing right now. The first path feels fast. The second one holds up every time.

Before anything else happens, before a single tool comes out, before a single assumption forms, you take pictures.

Photo the entire system in its environment. Make, model, nameplate, the unit itself, the surrounding area, the condition it is sitting in when you arrive. All of it. Before you touch anything.

No phone? Get one. No signal? Pen and paper. Get medieval if you have to. Documentation is not paperwork for the sake of paperwork. It is cover your ass in physical form. And in this trade your ass will need covering more than once.

Now. Assumptions.

You know what assume spells out when you split it up. If not take a moment and play word scramble with it. We will wait.

Your attitude, your personal history, your gut feeling about what this probably is, all of it stays in the truck. Window cracked so it can breathe. It will be there when you get back.

What you bring inside is a neutral state. Not chemically induced neutrality, that conversation is for a different book. Disciplined neutrality. The deliberate separation between what you have seen before and what is in front of you right now.

Because this is not before. This is now.

Experience gets loud under pressure. The owner is standing behind you close enough to leave a drool trail on your shoulder. Dispatch just sent a friendly note informing you that two escalated calls have been added to your already extended day. Panic starts creeping in and your fingers twitch toward that quarter inch nutdriver.

Stop.

I know that sounds counterintuitive. The pressure is real and it is not going anywhere. But that pressure is exactly why stopping matters. A technician who moves under pressure without discipline is not working faster. He is just making mistakes with more urgency.

Resist the urge to call out your diagnosis before you have confirmed anything. Maybe you are right. Maybe you nail it, problem disappears, you are out the door and on to the next one.

What happens when you are wrong?

You just contaminated your own diagnostic process with assumptions instead of facts. Now you are chasing a conclusion instead of reading a system.

Say you call it the capacitor. You are confident. You have seen this before. You replace it. Problem goes away. You leave.

Three weeks later the same call comes back. Because the capacitor was not the problem. The incoming amperage was. The capacitor was just overloaded, absorbing the consequence of a problem that was never located. You replaced a symptom and called it a fix. The system let you believe it because it had no choice.

Like your workday, it was just overloaded. Not broken. Overloaded.

Observation is the first real tool you pick up on any call. Eyes before hands. Experience informs the process. It does not dictate the outcome.

Let the system show you what it is doing. Then prove it.

That is where we are going next.

Chapter 5: Defining the Expected State

Before anything gets touched, one question gets answered first. What should this system be doing right now at this exact point in its sequence.

A specific, observable condition that can be confirmed or denied. Something the system can actually answer.

The expected state creates a boundary. The system either meets the condition or it does not. That binary is the entry point. The gap between what should be happening and what is happening is where the problem lives. You cannot find that gap without first defining one side of it.

This step gets skipped constantly. The system sits quiet, something feels off, and action starts immediately. A part gets checked because it looks suspicious. A component gets swapped because it failed on a similar job six months ago. Movement fills the space where process should be.

Defining the expected state slows that first move just enough to make it mean something. One question answered before anything else. What should be happening right now? That answer determines every move that follows.

Without it the work has momentum. Momentum is not the same thing as direction.

Expected state is simple but it is required for a full diagnosis. If you flip a light switch to on your expected state is illumination. That is as straightforward as it gets.

Until you see something that looks wrong. Until pattern recognition starts whispering in your ear. Somehow the expected state disappears from the process and experience takes the wheel. This is how seasoned technicians get burned. Figuratively and literally. Pay attention to this one.

Everyone wants to find and solve the problem fast. We pride ourselves on it. As time goes on we get genuinely good at it. Frighteningly so. I have watched technicians diagnose heating systems by observing smoke pattern from the chimney before they ever stepped foot inside the house.

As experience grows so does the ability to detect anomalies quickly. Are we right one hundred percent of the time? Only when we are arguing with our spouse.

Every technician has a different accuracy rating and it comes down to one question. Did you guess based on pattern or did you know based on evidence? Smoke coming out of a chimney or confirmed combustion data from inside the unit? There is a significant difference between those two starting points.

This is a crime scene. You are the detective.

Start with the expected state. When I engage this system what does the manufacturer say should happen. That is a fundamentally different question than what you think should happen based on something you saw online once.

Use the data you already gathered. Make, model, serial number, tonnage, equipment type. Whatever is in front of you has a manufacturer and that manufacturer documented exactly what this system is designed to do under normal operating conditions. That documentation is your expected state. Everything else is opinion.

The equipment is telling you its expected state was not met. That is why you are standing in front of it.

The manufacturer already answered the question before you arrived. Your job is to hold that answer in front of everything you do next and refuse to let experience talk you out of it.

Start there. Every time.

Chapter 6: Observing the Actual State

The expected state is defined. Now the system gets to respond.

That answer has to be specific, direct, and describable by anyone who walks up and looks without needing interpretation. "The fan is running." "The display is off." "The unit starts and shuts down after approximately three seconds." These hold their shape regardless of who is looking at them.

Observing the actual state means describing exactly what is happening without adding meaning to it. Watching the system versus explaining the system. One produces usable information. The other produces a story that may have nothing to do with what the system is actually doing.

Observation comes before explanation. That sequence is not flexible.

There is a habit that shows up here that destroys the process before it gains traction. The system does something unexpected and the brain moves to explain it before the behavior has been fully observed. The observation turns into a conclusion before it gets recorded.

Observe first. Completely. Without interference.

The system is already telling you what it is doing. The job at this step is to listen to it accurately.

Raise your hand if you are guilty of this.

The system starts its process and reaches a pattern you have seen before. For argument's sake a pressure switch failed to close. You know this. You have seen it before. The board is flashing a pressure switch failure code confirming everything your experience already told you.

Your brain lights up. Easy fix. Swap the pressure switch and you are back on the road. Might even have time to flirt with the gas station clerk on the way to the next call.

You tell the customer the problem is resolved. Just a pressure switch. You have one on the truck.

Back to the truck.

Here is where it matters. Careful selection of the replacement is required. The inches of water column must match manufacturing specifications exactly. This is not a suggestion. Failure to install the correct pressure switch increases the potential for serious harm to people and equipment. Read that again. Do not under any circumstances carry the mentality of close enough into this trade. It is not close enough. People are trusting you with their safety, their home, and their family. Respect that or find a different career.

Back inside. Switch installed. Unit runs. No problem.

Gas station clerk here you come. You are finishing the paperwork with the customer and the unit shuts down. Locks out completely.

Your customer is now looking at the bill. Looking at you. Looking at the bill.

"I have questions." They say.

And the technician who jumped on a pressure switch because the board told him to and his experience confirmed it does not have the answers to those questions. If you followed that, you are keeping up. Everyone else please use the exit on the right.

The process failed because a step was skipped. Pattern recognition substituted for observation. Experience moved faster than proof and the system had its own story running the entire time.

In this scenario the pressure switch failed because the homeowner stored his barbecue grill next to the intake pipe over the winter with a loose plastic cover on it. The unit ran until it consumed the available oxygen in the intake. The plastic tarp would get sucked into the intake forming a nice seal, then once the unit locked out the tarp would relax. Just like us, fire needs oxygen. Testing the pressure switch produced the readings it should have. By the time the technician arrived the unit was already in lockout. The board flashed the code. The technician saw the code, thought about a beef stick and a gas station clerk, and went to the truck for a pressure switch.

The system was running a completely different story based entirely in reality.

If you have not figured it out yet, I was that technician. I spent the rest of that day at that call cycling through pressure switch and board swaps that each worked for roughly one hour before the unit locked out again. One moment of discipline would have changed the entire day.

Stop. Think. Observe. Plan.

That sequence would have given me a hot coffee, a beef stick, and a moment of peace in the middle of the chaos. Instead I ate my pride in a customer's driveway until I found a barbecue grill cover wrapped around an intake pipe.

This industry can be brutal. Usually it is brutal in direct proportion to the steps you skip.

Let's keep moving so you are more prepared for your own chaos than I was for mine.

Chapter 7: Identifying Deviation

The expected state is defined. The actual state has been observed. Both sit in front of the process grounded in confirmed conditions. The next move is direct. Place them side by side and state the difference.

That difference is the deviation. A clear, nameable mismatch between what should be happening and what is happening at this exact moment. A stated, confirmable gap between two known conditions.

Until that mismatch is named the process has no fixed point to work from. Movement can happen. Parts can be handled. Time can disappear. None of it connects to anything because there is no defined target anchoring it.

Naming the deviation closes the open space. It draws a hard line around the problem and holds everything else outside it.

A defined deviation is a factual statement about a confirmed condition. It has to be grounded in what was observed.

Name it. State it clearly. Hold it.

Everything the process does from this point forward has to connect back to that named condition. If a step cannot be tied directly to the deviation it does not belong in the process.

Now here is where it gets interesting.

A career is made or broken on deviation. Because if your reality and the client's reality collide you better be standing on facts. In a commercial setting this becomes its own sport. Customer video recording your every move. Twenty different employees with twenty different versions of what is wrong and when it started. All of them delivered with full confidence before you have touched a single thing.

Most commercial customers only notice the equipment when it starts pushing unfavorable thermodynamics into their workspace. But the responses vary considerably once that happens.

Some office workers are soldiers. You know the one. Shawl on her shoulders, thermal hat, little electric heater humming under the desk. She adapted. She did not complain. She is surviving the conditions with dignity and zero drama. Respect that.

Then there is a different crowd entirely. They became experts before you arrived. Read that again.

By the time you reach the identifying deviation step you have already absorbed twenty versions of the problem from twenty different people. All of them are that person's reality. Hold onto that for a moment. Think about how hard people work to convince others that their version of events is the correct one.

Deviation removes all of that from the equation. Every version, every opinion, every confident diagnosis from the lady in the shawl or the facilities manager who Googled it on the way to work. None of it factors in. Deviation is what takes the bias out of the process entirely. That is what it does and that is why it matters.

So how do you use it?

With evidence. Because you are standing on observation and confirmed conditions. You are not asking the office staff what they think is wrong. You are not trying to reconcile twenty different versions of reality into one coherent diagnosis. You are confirming through your own observation what the system is actually doing versus what it is designed to do.

That is the difference between a technician who gets moved by the crowd and one who moves through it.

Knowing what is wrong is only useful if you know where in the system it is wrong. That is what comes next.

Chapter 8: Understanding System Sequence

Every system operates in steps. Each step depends on the one before it. A condition has to be present for the next condition to become possible. When that chain is understood the system becomes predictable.

When sequence is ignored every component in the system looks equally suspicious at the same time.

Sequence forces the system to be understood as a progression before any single part gets singled out. A component does not exist in isolation. It exists at a specific point in a specific order and its role in the failure cannot be evaluated without first understanding what comes before it and what depends on it.

The deviation is already named. Sequence places that deviation inside the operating order of the system. That placement is what turns a named problem into a located problem.

Place the deviation inside the sequence. The steps before it that completed are confirmed and cleared. The steps after it that were never reached are irrelevant. The work belongs at the boundary between the last confirmed step and the first step that failed.

Follow the order the system follows. Work at the edge where it stopped.

Sequence of operations. It even sounds beautiful. Because it is.

You will hear this term throughout your entire career. Get comfortable with it now because it is not going anywhere and neither are you until you understand it.

This chain of events is your trail through the system. Every machine has a designed path it follows when everything is working correctly. Your reality on the other hand looks like this. A set of schematics, a data plate, a manufacturing specifications chart, a frustrated customer, and dispatch leaving you a friendly reminder that you have ten more calls and the office is closing for the night.

You have heard twenty versions of the problem. You have been told to chase ghosts. You missed lunch entirely. It is snowing outside. You are now sitting in front of a machine that seems personally invested in questioning your life choices.

Discipline is needed once again. Tissues are in your pocket. You can process your feelings in silence on the way home.

The sequence of operations is the machine's discipline. Your job is to match it.

That starts with schematic and blueprint reading skills. These are not skills that develop in a day. They take time, repetition, and deliberate study. But they are not optional. A full understanding of what should be energized and the timing in which it should be energized is the foundation everything else builds on.

Schematics and electrical theory go hand in hand. Learn both.

Take a residential furnace as an example. A sequence of operations may look like this, depending on the equipment in front of you. That last part was added for that guy. You know who you are.

The thermostat calls for heat. Inducer motor energizes. Pressure switch closes. Low voltage travels through the rollout switches. The 24 volt safety circuit completes. Relay closes allowing voltage to the hot surface ignitor.

Timing sequence occurs. Gas valve energizes. Ignition takes place. Flame rectification is verified. Blower motor establishes.

That is a sequence of operations. You just read one.

If you are looking at an air conditioner and expecting fire because that is what the last unit called for, there is a special place in the universe reserved for that decision. No further comment.

If you understand schematics and can read data plate information accurately, sequence of operations is where you start locating problems instead of chasing them. Every minute spent wandering outside the sequence is another call aging on the board. The numbers keep climbing the longer you linger. Welcome to HVAC reality.

You observed the deviation. You know where the sequence broke down. Now the job is proving exactly where and why. Not guessing based on what the deviation suggests. Proving it.

The sequence showed you where to stand. The next step proves what is actually there.

Chapter 9: Cause vs. Effect

The system presents what is easy to see. Water on the floor. A stain on the ceiling. A noise at a specific point in the cycle. A smell under load. These conditions announce themselves. They pull focus immediately. They feel like the problem because they are the most obvious thing happening.

That visible condition is the effect.

The cause sits somewhere else. Upstream of what is being seen. Operating in a location the effect has been pointing away from the entire time. Confusing the two sends the process in the wrong direction with full confidence. That is a more expensive mistake than being lost without direction because it takes longer to recognize.

The effect is still useful. It shows where the system is expressing the problem. It provides a position. It narrows the search area by showing where the result ended up which points back toward where the result originated. The effect is a starting point for tracing backward not a destination for working forward.

The cause must create the effect. The effect does not create itself and cannot resolve itself regardless of how thoroughly it gets addressed.

Find the cause. Trace the effect backward to where it originates.

Within the sequence of operations the chain was broken somewhere. That break created the cause and effect scenario now sitting in front of you. Effects speak louder than causes most of the time and they will catch you slacking if you let them.

Do not let them. Focus.

Here is a call that makes the point clearly.

Customer complaint. Water is dripping from the ductwork in her new home with her new furnace.

The effect. A hostile customer, now furious, with puddles of water throughout her home.

The cause. Unknown. Start there.

Run the equipment through its sequence of operations. Verify the customer's claim. The effect is real. Water is coming out of the plenum. That part is confirmed.

The cause is not the furnace. Water obeys gravity and this furnace happens to sit lower than the source. The sequence of operations on the equipment checks out clean. Nothing in the system is producing water that can defy gravity, without a pump. Which means something above it is the cause.

Now there is a second effect to manage. A further upset homeowner when you explain that the water she is watching fall is not coming from her furnace. Her reality is that it is the furnace. It is new, it is the only thing that changed, and it is the obvious suspect.

Removing a section of ductwork to show her the water traveling down from above helps align both realities. Remember what was covered under deviation. Evidence over opinion. Hers and yours.

What was found upstream was a five gallon water jug with a slide out dispenser. She had stored it in a cabinet directly above a supply register. When she put it away it cracked open slightly. Water followed gravity down through the cabinet, into the supply, and out through the plenum below.

Effect. Water.

Cause. An honest mistake with a water jug.

Sequence of operations on the furnace. Safe and sound. Nothing to see here.

The effect is what you and the customer experience together. The cause is what you were sent to identify. Keep those two things straight and you will solve problems other technicians walk away from scratching their heads.

That distinction matters in every call. It matters even more in the next step when you start using what you know to locate exactly where the system broke down.

Chapter 10: Using Verification Points

The sequence is in place. The deviation has a name. The process is standing at the right place.

But knowing where to stand is not the same as confirming what is there. That confirmation requires something more deliberate. Each step in the sequence has to be checked under the current condition before the process moves past it.

That is what verification points are. They are the moments where the process stops moving forward and proves what it is standing on.

Without them sequence becomes an outline instead of a method.

Verification keeps every step grounded in something confirmed. It converts the sequence from a map into a proof. Each point either holds or it does not. A condition is present or it is not. That binary is what gives the process its direction.

Before a verification point means anything the tool or method being used to confirm it has to be trusted. A reading from a faulty meter is not verification. A visual check on a component that requires measurement to evaluate is not verification. The confirmation has to be tied to something that is itself reliable.

Confirm before acting. Not after. Not while. Before.

Each confirmed step removes part of the system from consideration permanently. Each failed step becomes the focus immediately.

The sequence showed where to stand. Verification proves what is actually there.

Skipping steps in this area should be listed as an international crime. And to be clear, I am just as guilty as anyone else.

As a technician you will feel pressure equivalent to pushing out diamonds before breakfast. It will not be in your intestines. It will sit squarely on your shoulders and whisper failures directly into your ear.

Resistance feels futile. It is not.

It feels faster to look at a system, see a fan motor not running, swap it out and leave. You get home earlier. The company is happy. The customer seems happy.

Until they are not.

Until the callback arrives and you realize it was never the fan motor at all. It was a start capacitor that decided to go on vacation. You swapped a fan without verifying the capacitor because you were pretty sure.

Pretty sure just cost you a conversation where a three hundred dollar motor becomes a fifty dollar part. It also costs you the time to swap the old fan back out or hand the customer a new one because a used fan cannot be resold. Your boss starts doing math. Dollars out the door versus dollars you bring in. That math is real. Ask anyone wearing a tie.

Verification starts at the beginning of the sequence of operations. Not at the effect.

Do I need to say that louder for the technicians still sitting in the truck?

Verification starts at the beginning of the sequence. No skipped steps. No assumptions. It is not optional and it is not flexible.

Here is what that looks like in practice.

Start at the transformer. Verify voltage, both high and low side. Proper voltage present, move forward. No voltage present, you just found your starting point and you did it in the first step.

After voltage is confirmed at the transformer, follow it into the sequence.

24 volts travels from the transformer to R on the circuit board. Is 24 volts present at R? Yes, move forward. No, stop. You already proved voltage exists at the transformer. Now prove why it is not reaching R. Run a continuity check between those two points. The schematic shows 24 volts should be present at R. If it is not, find out why. Is the wire broken between point A and point B? Check it. You own a meter. Use it.

R is confirmed energized. Now follow it.

R travels to the thermostat. Go upstairs. Measure R on the thermostat to ground. Is 24 volts present? That is the next confirmed step in the sequence according to the data in front of you.

R carries 24 volts and waits for an event. A temperature drop. That drop brings two contacts together and gives 24 volts a bridge into the W circuit. The W circuit sits there waiting to become a path for current. Doing nothing. I am almost jealous of the W circuit, works only half the year has the entire summer off. No voltage is present here, W is a path in the sequence of operation for 24 volts waiting on R. Temperature drops, condition satisfied.

W is now energized. You proved it. With a meter and a brief moment of wishing you were a circuit.

Head back downstairs. Pause on the way and let the homeowner know you are on it. Pat the dog. Back to the dungeon.

Here is what just happened. You confirmed voltage at the transformer. You confirmed the wire delivering that voltage to R. You confirmed R is energized and that the thermostat is completing the W circuit. Those specific points are cleared. The board itself has not been fully tested beyond voltage arriving at W. Keep that in mind as the sequence continues. You could have tested just W at the board. 24 volts present, yes or no would answer is the circuit complete.

What if the homeowner just tied R and W together upstairs? Always inspect each system component.

Now consider a different scenario. You arrive and find zero volts at the transformer.

A failed transformer is an effect. Something caused it to fail. Replace it without finding that cause and you are handing the next transformer the same fate.

What convinced this transformer to pick up a habit like smoking? That question matters before anything gets replaced.

Right here you have a choice. Not my circus, not my monkeys. Replace the transformer and move on. Or pull out the meter, fire up the grey matter between your ears, identify the verification points through the sequence of operations, roll up your sleeves, and find out why.

That answer lives in the break point.

Which happens to be exactly where we are going next.
Lucky you.

Chapter 11: Locating the Break Point

The sequence has been followed. Verification points have been confirmed one by one. Some steps held. One step failed. The process is now standing at the exact location where the system stopped moving forward.

That location is the break point.

It sits between the last confirmed working condition and the first confirmed failure. It is not a theory. It is a boundary defined by what the system itself has demonstrated. Once it is found the search area collapses from the entire system down to a single position.

The work happens at the boundary and nowhere else.

Follow the system forward from the last confirmed working point. Check each position in sequence. The moment the condition changes from working to not working that transition is the break point.

Find where it stops. Work there.

The break point is not a tennis reference. In this book the break point describes exactly what it sounds like. The precise moment in the sequence where the system stopped doing what it was designed to do.

Voltage in, no voltage out. Or voltage in and no desired effect. Notice the word effect was used there and not cause. That distinction still matters here.

At this point in the sequence of operations you have proven each component met the requirements for operation. Except this one. Every step before it confirmed. This one failed.

Swap the part and go home right?

Wrong.

Having evidence that a part is faulty is only half the job. The other half is understanding why it is faulty. Because if the why does not get answered the break point does not get resolved. It gets postponed. You will be back. Or someone else will be. Either way the problem continues until the cause behind the failed part gets identified.

Unless the part failed on its own. But that has to be proven too.

A bad inducer motor is not a conclusion. It is an observation. Bad inducer motor does not tell the client or the office how you arrived at that finding. More on documentation later. For now the target is understanding why that inducer motor failed.

Many things cause motor failure. Excessive heat is one of them. Have you done a temperature rise calculation and matched it against manufacturer specifications? Do you know whether heat is a factor in this failure or is that a hunch? Prove it. Do the steps required to isolate that variable.

Was it moisture in the basement driving up amperage through friction? Prove it. Pull an amp draw on the motor if it runs at all. If it does not run, open it up. Exploratory surgery right there at the scene. Test the windings. Do they ohm out? Does the motor shaft spin freely? Why or why not?

Here is what that process looks like in practice.

The inducer is the suspect. The motor is humming. Amps are climbing. The motor is tripping on overload protection exactly as it is designed to do. You have used the process correctly and science has cooperated. Eureka.

Then you open it up and find a dead mouse wedged between the fan blade and the housing.

Nobody saw that coming. Especially the mouse.

You would never have identified that cause by guessing at the motor from the outside. The part had to be isolated from the system and examined directly. That isolation is what revealed the actual cause behind the break point.

Controlled testing and isolation work together at this stage. You proved the part was at fault through verification. You isolated the part to find out why it was at fault. Both steps were necessary. Neither one was optional.

And none of it was possible without following every step that came before it.

The verification pointed you to the break point. The break point gave you a location. Isolation gave you the answer inside that location.

That is the process working exactly as it is designed to work.

Chapter 12: Confirmation of Cause

The fix has been applied. The part has been replaced, the obstruction removed, and the failed component addressed. The system responded. Something that was not working is now working.

The job is not finished.

A system that responded to a fix has not demonstrated that it is operating correctly. It has demonstrated that it responded to the fix. Those are not the same thing.

Confirmation requires the system to be run the way it is designed to run. Not under simplified test conditions. Not through a single cycle. Under real operating conditions, through its full sequence, from start to finish, until the behavior is stable, consistent, and matches the expected state at every point along the way.

That standard is not satisfied by a single response to a single action.

Oh do I know that feeling. The system kicks on, everything sounds right, and every instinct you have is telling you to get out of there before it changes its mind. High tail it out like the whole thing was a bad dream and hope it stays fixed.

What stops it from becoming a bad dream is discipline. One more time.

Repairing the system is not the same as confirming it. You may or may not have won. Time is the ultimate scorekeeper and it does not care how confident you felt walking out the door.

Some of you are rolling your eyes right now. Keep reading.

Run the equipment through a full cycle under load before you leave. Take the time. Do your paperwork in the mechanical room while the system runs. Keep your eyes and ears on it. Temperature probes in supply and return air, visible and monitored. Listen for squeaks, rattles, and anything that does not belong in the timing sequence. Run that equipment like it is yours and you know you are going to need it tonight.

You would be surprised what gets missed when the focus was locked on an effect. Time has a way of surfacing what the urgency of the call buried.

Keep an amp meter on the system. Monitor it like you are playing doctor and this is your patient. Is it breathing? How is the pulse? Are the amps stable? Is the temperature rise within the range the manufacturer specified?

Then something catches your attention.

The readings are drifting. Ten degrees outside manufacturer specifications and climbing. The longer it runs the further out of tolerance it goes.

Now the question arrives. Is this connected to the failed part that started all of this or is this a separate condition that was hiding behind the original problem and letting it take the heat?

Pun intended.

That question needs an answer before you close out. You have a choice. Answer it now, while you are standing in front of the equipment with your tools in hand and the system running. Or carry it home with you. Let it sit rent free in your head through the next few days until that fateful day becomes inevitable and dispatch hands you a return ticket with a highlighted note at the bottom.

Please speak with management after this call is completed. Thank you and have a nice day.

Answer it now.

Run the full cycle. Watch every transition. Compare every step to the expected operation. When the system holds across repeated cycles without deviation the cause is confirmed and the work is finished.

When it does not hold the process is not finished and the job does not close.

The system ran. It stayed finished.

That is confirmation. That is where the process ends. That is how you prove it.

Chapter 13: Documentation

I warned you I was coming back to this.

Documentation sucks. Plain and simple. Any technician in this trade will tell you they hate taking the time for it. It slows you down. You have to explain the same findings six different ways across six different forms. The customer is inside tapping his fingers wondering if you are watching cartoons in the truck.

Meanwhile you are out there putting your findings into a legible format.

Take note of that last word. Legible.

Anything you write for business purposes gets written in block print. Cursive and sloppy handwriting are not style choices. These documents may end up in court.

Pause. Did that register?

The document you are creating right now in that truck can either save you in a legal dispute or cost you everything because of shortcuts and poor discipline. You will pay for it one way or the other. The question is whether at some point in your career you stop and diagnose the technician doing the diagnosing.

Self reflection is not optional in this trade. It is part of the job.

The reasoning behind your statements and the evidence you present cannot come from a gut feeling or a pattern you half remember. Facts only. With description. With every photograph you took of the job site to prove your diagnosis is solid. Your documentation must be clear enough that anyone following your work can read it and understand exactly what you found and exactly how you found it.

As mentioned earlier, bad inducer motor is not a diagnostic.

What happens when the client disputes your bad inducer claim? What happens when Billy Bob’s Plumbing and Heating disputes it? What happens when you are defending that diagnosis a year later in a courtroom? Any idea how many bad inducers you have touched between that call and that court date?

Neither do I. But if you are in this business it is quite a few.

Now look at what Billy Bob cannot dispute. Because you proved it.

Facts. Scientific facts. No personal opinions. No hunches dressed up in technical language. Facts and nothing but.

Here is what that looks like on a work ticket.

Arrival time: 4:15 pm

Complaint: No heat

Diagnostic: Checked system sequence of operations against manufacturing data. Sequence failed. Identified an inducer motor failure. Circuit board code E506 displayed confirming pressure switch did not close. Investigation of said inducer motor revealed a rodent impacting operation. Rodent removed and inducer isolated for further diagnostic.

Detached from the system the inducer motor runs at 0.02 amps. Reattached to the system under load the inducer motor runs at 0.06 amps.

Currently the system is operational. However replacement of the motor is advised to ensure reliability. Two conditions have impacted this inducer motor.

Windings have been subjected to temperatures outside of normal operations. The motor overload has been exercised repeatedly.

Both are concerns revolving around the future reliability of this part. To be clear this part is functional, however it has been abused and that directly affects long term reliability.

Air flow fins inspected and appear intact however microfractures not visible to the eye cannot be ruled out.

Recommended: Replacement of inducer motor and housing. Motor and housing replacement ensures equipment is returned to manufacturing specifications and reliability is restored.

End time: 5:21 pm

That is how you write a work ticket. Clear. Factual. Defensible. A judge can read it. A customer can read it.

Another technician can read it and follow exactly what was done and why.

I started this book telling you to do your own homework. Closing a call properly, documenting your steps, and expressing your findings in language that holds up under scrutiny is part of that homework. Nobody is going to teach you this part. Most companies hand you a tablet and wish you luck.

This section exists to give you the foundation to stand on your career instead of under it.

You wear many hats as a technician. Expressing yourself professionally and factually is one of them. It is not the most glamorous hat. It is the one that keeps you employed, keeps you out of court, and keeps your name attached to work you can defend.

Documentation is the last thing you want to do and the first thing that saves you. Treat it accordingly.

Chapter 14: Mr. Smith

Oh you want to see the method working?

Well that's for you to apply in your reality. This method is not the only way. It most certainly can be improved upon, by you. Make it your method and then some.

I can tell you a reality I once had that tested the method I preach. And I will prove it.

I was a lead senior technician for a very large, very reputable company. My company had a problem and it wasn't small.

Multiple service technicians had been sent to handle this customer's complaints. None of them had a diagnosis. All of the service reports read as educated guesses. Multiple companies had been to this customer's house, all had failed to resolve the issue, including three of our own technicians.

The biggest problem arrives in the form of a brand new furnace. If you guessed that a tech talked the homeowner into a replacement and made guarantees that the problems would be gone, your experience is talking. And it happens to be right.

We will call him Mr. Smith. Frankly I don't recall the real customer's name anyway.

Mr. Smith called management hot. His new furnace was still doing the same thing the old one was. My orders came simple from the top.

"We don't care if you have to sleep at the customer's house. You are paid to deal with these problems. Don't come back to us with more problems. Bring us the solution and prove it."

Well okay then. Walk around the truck, do the inspection. Off to hell I go.

Before I left the yard, photos. Make, model, dispatch notes, everything documented before boots hit the ground. That habit starts before you arrive and it does not stop until the call closes.

On the way to the customer's house I thought over the complaint.

Intermittent heat. Customer states unit runs fine all night, into the next day. When he leaves for work, furnace runs no problem. When arriving home, furnace seems to stop working and goes into hard lockout. Customer states he has to shut the emergency switch off and back on and the system will then operate normally for another twenty four hours. Customer states on weekends the system runs fine with only occasional problems, seemingly as the weather warms up. Customer is hostile. Be advised he has had multiple companies including ours.

Two systems, old and new, produced the same effect. That pointed toward an external influence, but it still needed to be proven. The equipment was not the cause. The environment was the suspect. That is where the expected state gets defined, not at the component level but at the operational level. What should be happening and what is the system actually doing.

Pulling into his driveway he was already outside, arms crossed, with that look of "you better be a magician" written all over his face.

Careful introduction. I like to add a touch of humor for the sharp. Still keep it professional. And I cannot stress this enough. Read the room. Customer is already hostile. Yes sir. No sir. Yes ma'am. No ma'am. That is the designated vocabulary with a pissed off customer. Period.

It's not about you. Sometimes the customer just needs to vent, be heard, and know someone is listening.

That's what went on here for a considerable amount of time. My mouth was shut and this customer blasted me and everything about my trade. We are all idiots. He was absolutely right too. But you are also representing your company and your trade. Both worlds must exist here with respect.

I validated his anger. Then I told him my theory.

I believed he was doing something without knowing it that was influencing this problem.

He denied this like I had just put him on trial for the death row of people who don't change their filters.

After patiently waiting through that explosion, I asked him if he would go through his routine when he came home from work. Begrudgingly he agreed.

The first thing he did when we entered his home was take me directly to the new furnace we installed. Experience would already tell you what that means. But this is the customer's home, so I followed and listened to every complaint about the installation.

"Did you know one man on your team smokes? It's disgusting. In and out of my house smelling horrible. Who's going to pay for that?"

Normally you could redirect. Time is hourly, I don't want to run the bill up, let me get to work. But on callbacks the dollar amount is often zero. So what do you do? The customer has you as a personal punching bag and the company is watching the minutes tick by like you are personally responsible for this mess.

In a lead position, you are. You own it.

"My company gave me authorization to sleep here next to the furnace if I have to. I'd rather not. Bad back, cranky wife, screaming kids, and an ungrateful cat await. On second thought, maybe I will stay. I'm getting paid after all. Your basement is very nice."

"What do you need me to do?" he asked.

"Your routine. From walking in the front door to every motion in between that's repetitive. Going to a certain room first, the kitchen, sitting down to take off your boots. Anything done daily after work. Because I think something you touch is causing this problem. So please, touch everything."

This is observation of actual state. Not the system's state. The environment's state. What is actually happening in this space that the service reports never captured because nobody stayed long enough to watch.

For context, this was an eighty percent builder model furnace. Very simple design. Fault codes were flashing lights in a number sequence. His old furnace didn't even have that.

I also noticed two step-down transformers sitting in the corner of the basement. The customer said they belonged to the old furnace. A mini graveyard. Somebody had been replacing them and moving on without asking why they kept failing. That detail went in the back pocket. Effects were being treated. The cause was never touched. That is the difference between closing a call and solving a problem.

When the customer reached the slop sink in his mudroom, I heard the furnace click and lock out. The code that flashed was a short.

I grabbed the customer and verified. This was the effect he had been witnessing. He confirmed it, after removing partial use of my left ear.

Now we had a deviation. Defined, witnessed, and confirmed by both of us in real time.

Keeping cool, I disconnected the thermostat wires from the furnace and jumped R and W after a reset. Furnace running. Thermostat wires removed. Customer ready to see something happen.

This is where sequence of operations earns its place. Isolate the circuit. Remove the variable. Prove what is carrying the problem before you go looking for where it lives.

A continuity test performed. I explained to the customer that no tone is good.

"Go back to the slop sink."

Bam. Solid tone.

Verification point confirmed. The thermostat circuit was the break point. Now follow it to the cause.

Following the thermostat wires with the customer hot on my heels, I found an area under the slop sink where the floorboard sagged every time he stood in that spot. Whoever installed the original ductwork had notched a section of the floor joist and run ductwork through it. And because presentation matters apparently, they ran the thermostat wires on top of the galvanized steel.

Over years of standing in that spot the wire insulation had worn off. Bare copper exposed. With enough weight from above, those wires shorted directly onto the ductwork. Sometimes briefly. Sometimes long enough to blow fuses, have technicians come out, bypass the fuse to get the customer by, and leave. Hence the graveyard of transformers in the corner.

That was the cause. Everything else, the lockouts, the codes, the transformers, the brand-new furnace, those were effects. Every technician before me treated an effect and handed the next one the same problem with a new label on it.

Years of problems. Multiple companies. A brand-new furnace.

A snip and a wire nut.

When the customer understood what he was looking at he told me to leave his home. I knew walking away that it wasn't me he was angry with.

That man had just lost all faith in the HVAC trade.

Closing Statement

You either ran the process, or you did not.

There is no partial credit in a mechanical room. The system does not grade on effort. It runs or it does not. It holds or it does not. You proved it or you are guessing and calling it experience.

Most technicians who read something like this will nod along and change nothing. That is not a prediction. That is a documented pattern across every trade, every method, every book written by someone who bled for the information they put on the page.

You are not most technicians. Or you are. That gets settled in the field, not here.

This method works when applied without compromise. Every shortcut in it costs you something. Sometimes it costs you a callback. Sometimes it costs you a customer. Sometimes it costs the next technician who inherits the system you left behind. The bill always gets paid. The only variable is whose name is on it.

Diagnose yourself the same way you diagnose a system. No assumptions. No patterns substituting for proof. No comfortable conclusions. What should you be doing. What are you doing. Name the gap. Work there.

That is the whole job.

There are moments in this career where you stop and just laugh. Because if you do not, you are going to cry. Keep a sense of humor. Be as weird as you want on your own time. That keeps you sane.

A bond unspoken amongst men of the trade. We look out for one another, we laugh together, we work together and we stay open minded of each other's thought process. We bond together because we grow together and that growth turns into loyalty to the brothers in trade.

Once I was tasked with a clear objective straight from the top. Direct your team to fill water bottles with their customers tap water and put the customers' address and phone number on the bottle. For every bottle you fill and tag, this water company will give you five dollars per bottle. My job was to deliver the news. Enforcement was above me at the time. I turn wrenches.

Well of course there were questions from a bunch of thinkers. I mean hello.

First question: What makes this company think I am going to invade my customers' privacy for five dollars and give away their information so they can try to sell them water purifiers?

Second question: Have you lost your mind?

Third statement: Oh, we will fill them alright. Give them here. It won't be the water they are expecting though.

I just deliver the message boys.

By the end of the week, I was approached by the man in the tie and he was not happy. Every sample sent back was urine and he was particularly upset about the references implied on the labels.

I laugh now. But that was a hot situation.

Now get out of my truck.

OPERATIONAL CHECKLIST

STOP

Stop. Think. Observe. Plan. Before anything else.

TAKE PHOTOS

Document the system in its environment. Make, model, nameplate, condition on arrival. Before you touch anything.

DEFINE EXPECTED STATE

What should this system be doing right now at this exact point in its sequence? Use manufacturer data. Not memory. Not pattern. Data.

OBSERVE ACTUAL STATE

What is the system doing right now. Facts only. No interpretation. No conclusion. Record what is present and what is absent.

IDENTIFY DEVIATION

Place expected and actual side by side. State the mismatch clearly and specifically. This is the target. Everything connects back to it.

ESTABLISH SEQUENCE OF OPERATIONS

What is the correct order of operation for this specific system. Place the deviation inside that order. Confirmed steps behind you. Unconfirmed steps ahead of you.

USE VERIFICATION POINTS

Confirm each step in sequence before moving past it. Confirm before acting. A condition is present or it is not. Proof replaces assumption at every point.

LOCATE THE BREAK POINT

Where does confirm working stop and confirmed failure begin. That boundary is the address. Work there and nowhere else.

SEPARATE CAUSE FROM EFFECT

What is the system expressing and what is creating it. Trace the effect backward to its origin. Address the cause.

CONFIRM THE FIX

Run the system through its full cycle under load. Monitor amps, temperature, timing sequence. Watch every transition. When the system holds across repeated cycles without deviation the job is finished.

DOCUMENT

Block print. Facts only. Photos attached. Defensible by anyone who reads it including a judge.

RAPID USE DIAGNOSTIC FLOW

STOP. Think. Observe. Plan.

Take photos before touching anything.

What should be happening right now?

What is happening right now?

Where do those two stop matching?

That mismatch is the target.

Where is the system in its sequence of operations?

What must happen at this step?

Is that condition present?

YES. Confirmed. Move forward.

NO. This is the focus. Work here.

Verify before acting. Confirm the condition. No assumptions. No patterns. Evidence only.

Change one variable. Observe the result directly.

Did the system respond correctly?

YES. Continue forward through the sequence.

NO. Reassess within the boundary. Do not expand the search.

Is the condition you are addressing a cause or an effect?

Trace backward to the source.

Apply the fix. Run the full cycle under load. Monitor amps, temperature, timing sequence.

Does the system hold across repeated cycles without deviation?

YES. Document. Close the call.

NO. The process is not finished. The job does not close.

The system ran. It stayed finished.

That is how you prove it.

About the Author

Sgt. Cognition is an independent author and publisher operating under the imprint Cognition Admission. He brings over twenty years of HVAC/R and facilities leadership experience to the page, most of his time spent as a lead senior service technician diagnosing systems across residential and commercial environments.

His catalog spans multiple divisions including Code: Primal, Code: Feral, and Code: Faith.

www.sgtcognition.com

Made in United States
North Haven, CT
04 November 2026

21767535R00044